Food
51

带刺的荨麻
Stinging Nettles

Gunter Pauli

冈特·鲍利 著

田烁 译

丛书编委会

主　任：贾　峰

副主任：何家振　郑立明

委　员：牛玲娟　李原原　李曙东　吴建民　彭　勇
　　　　冯　缨　靳增江

丛书出版委员会

主　任：段学俭

副主任：匡志强　张　蓉

成　员：叶　刚　李晓梅　魏　来　徐雅清　田振军
　　　　蔡雩奇

特别感谢以下热心人士对译稿润色工作的支持：

姜竹青　韩　笑　杨　爽　周依奇　于　哲　阳平坚
李雪红　汪　楠　单　威　查振旺　李海红　姚爱静
朱　国　彭　江　于洪英　隋淑光　严　岷

目录

带刺的荨麻	4
你知道吗?	22
想一想	26
自己动手!	27
学科知识	28
情感智慧	29
艺术	29
思维拓展	30
动手能力	30
故事灵感来自	31

Contents

Stinging Nettles	4
Did you know?	22
Think about it	26
Do it yourself!	27
Academic Knowledge	28
Emotional Intelligence	29
The Arts	29
Systems: Making the Connections	30
Capacity to Implement	30
This fable is inspired by	31

一只瓢虫带着她的孩子穿越森林。他们从一株植物飞到另一株上，吃着植物上的蚜虫。瓢虫一抬头，刚好看到一只小山羊伸着舌头，馋馋地盯着那些茂盛的绿色荨麻。

"孩子，你饿了吗？"瓢虫问。

A ladybird guides her young through the forest. They move from one plant to the next, feeding on aphids. She looks up and sees a young goat with his tongue sticking out, hungrily looking at the lush green nettles.

"Are you hungry, kid?" she asks.

"孩子,你饿了吗?"

"Are you hungry, kid?"

从来不知道被荨麻刺扎到会这么疼

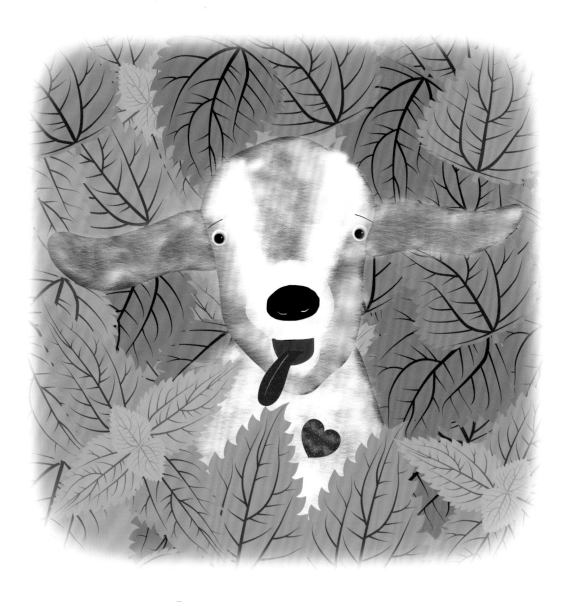

Never knew nettles sting so badly

"我实在是太饿了。我想吃这些荨麻,可是舌头却被荨麻的刺扎疼了,你看,我舌头上到处都是荨麻刺扎出的小红泡了。"山羊哀叹。

"你妈妈没有告诉过你要小心一些吗?"

"她当然说过啊,但我从来不知道被荨麻刺扎到会这么疼。"

"I am so very hungry. I tried these nettles and now my tongue is itching, and it has lots of red bumps all over," laments the goat.

"Your mom never told you to be careful?"

"Of course she did, but I never knew nettles could sting so badly."

"荨麻是我们这些昆虫最好的避难所了,因为荨麻能保护我们不被奶牛或山羊吃掉。你的舌头被刺痛一次,以后就不会再来这里吃了。"瓢虫笑着说。

"但是,我妈妈告诉过我,这些荨麻有助于治疗疾病,比如关节炎和风湿病。"

"Well, nettle fields are a great refuge for us insects, as there is little danger that we will end up in a cow or a goat's belly. Your tongue will only hurt once. After that you'll never come back for more," laughs the ladybird.

"But Mom did tell me these stings are good for treating health problems, like arthritis and rheumatism."

……我们昆虫最好的避难所……

...great refuge for us insects...

妈妈告诉我荨麻富含铁

Mom told me, nettles are rich in iron

"你妈妈说得很对。但是,首先,你这么年轻,不会得这些病的;其次,我从来没听说过舌头会得风湿病。"瓢虫咯咯地笑着。

"妈妈还告诉我,荨麻富含铁,铁是我身体里造血必需的元素。"

"She is absolutely right. But, firstly, you are too young to suffer from these illnesses, and secondly, I don't know of any animal with rheumatism of the tongue," chuckles the ladybird.

"Mom also told me nettles are rich in iron, and that is what I need to make lots of blood."

"你有一个非常有智慧的妈妈。"瓢虫点头说,"你为什么不做荨麻汤喝,而不是生吃它们呢?"

"我还没有学会怎样做汤呢。汤好像是人类的食物,不是我们动物的。"

"You have a very wise mother," says the ladybird, nodding. "Why don't you make a soup of the nettles, instead of eating them raw?"

"I haven't learned how to make soup yet. It seems more like something people eat, not animals."

做荨麻汤

Make soup of the nettles

他们的衣服不扎人吗?

Didn't their clothes sting?

"人类利用荨麻有很久的历史了。"瓢虫认真地说,"他们的王室都曾经穿过荨麻做的衣服呢。"

"啊,那他们的衣服不扎人吗?"小山羊惊讶地睁大了眼睛。

"People have been using nettles throughout history," muses the ladybird. "Their royals even used to be dressed in it."

"But didn't their clothes sting?" The kid's eyes go wide with surprise.

"当然不!他们会先煮荨麻,然后用荨麻纤维来制作漂亮的服饰。荨麻制成的衣服非常结实,过一百多年都不会烂。"

"好神奇啊!"小山羊喊道,"那就是说,爷爷买了一件荨麻制成的衣服,他的孙子的孙子的孙子都还可以穿咯!不过,我猜等他的后代穿的时候,他的衣服一定太过时了。"

"No, of course not. They would first boil it and then use the fibres to make beautiful dresses and cloaks. The clothes were so strong, they would last for a hundred years."

"That's amazing!" cries the kid. "So the clothes a granddad bought would still be dressing his great-great-grandchildren. But I'm afraid his clothing must have been out of fashion by the time they got to wear it."

用荨麻纤维来制作漂亮的服饰

Fibres to make beautiful dresses and cloaks

为什么人们要穿纯棉的衣服

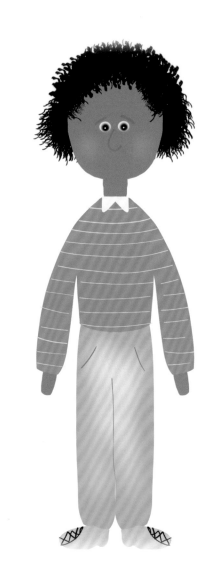

Why do people dress up in cotton

瓢虫笑了。"是的,过去,祖辈留给下一代的衣服是一种特殊的礼物,就像日本妇女那样,她们现在仍然将自己的和服留传给后代,那些和服有几百年的历史了。和服是永远不会过时的。"

"真的吗?那为什么人们要穿纯棉的衣服或石油产品制成的衣服呢?石油这种资源可是不可持续的。"

The ladybird smiles. "Yes, in the old days receiving clothing was a special gift from one generation to the next, just like Japanese women who still hand down their kimonos, hundreds of years old. Kimonos never go out of fashion."

"Really? So why do people dress themselves up in cotton, or in clothing made from petroleum, a resource that can't last?"

"嗯，聪明的小朋友，原谅他们不懂得珍惜现有资源的行为吧！我可不会用我的荨麻来做任何交换的。荨麻不仅给我提供食物和避难所，而且还是一个很好的邻居呢，荨麻丛里到处是漫天飞舞的蝴蝶。"

……这仅仅是开始！……

"Well, my clever young friend, they should be forgiven for not appreciating what they have. I will not trade my nettle field for anything. Not only does it give me food and shelter, it is also a wonderful neighbourhood to live in, one so full of butterflies."

... AND IT HAS ONLY JUST BEGUN!...

……这仅仅是开始!……

...AND IT HAS ONLY JUST BEGUN!...

Did You Know?

你知道吗？

Nettles have been used for centuries in natural medicine to reduce inflammation and treat allergies.

人类使用荨麻作为缓解炎症、治疗过敏的天然药物已经有几个世纪的历史了。

Nettles taste like spinach. They can be brewed as a tea, blended as a pesto and made into a purée. Nettle soup is common in eastern Europe. In the UK and the Netherlands it is a popular ingredient in cheese making.

荨麻的味道像菠菜。荨麻可以用于制茶，也可以混合成香蒜酱，做成果泥食用。荨麻汤在东欧国家很常见。在英国和荷兰，荨麻经常用于制作奶酪。

𝒩ettles have been used for clothing, tablecloths, bedsheets, sandbags, rucksacks and even parachutes. The national dress (gho) of Bhutanese men is made from nettle.

荨麻已经用于制作衣服、桌布、床单、沙袋、帆布包，甚至降落伞。不丹男子的传统服饰（帼）也是荨麻制品。

𝒩ettles are perennials and need no irrigation or pesticides. They grow abundantly and can be harvested from Nature; there is no need to plant it as a crop. People are ignorant about the history and the use of nettles and therefore classify this plant as weed.

荨麻是多年生植物，不需要灌溉或杀虫剂。它们生命力顽强，靠自然生长就可收获。没必要像种植庄稼那样来种植荨麻。人们由于对荨麻的历史和用途缺乏认识，而错误地将其当作野草。

The nettle's sting was developed as a defence against grazing animals. With the rare exception of hungry goats and sheep, grazers will leave the nettles alone, creating a protected area for insects and plants alike.

荨麻的刺是为了抵御食草动物而慢慢进化出来的。除了偶尔有些饥饿的山羊和绵羊之外，食草动物们一般都不会去碰荨麻，这就为那些昆虫和类似的植物留下了生存空间。

Ladybirds are beneficial predators living on plants while feeding on mites and aphids. During its lifetime of 3 to 6 weeks each ladybug eats approximately 5 000 aphids.

瓢虫是寄生在植物上的食肉动物，以螨虫和蚜虫为食。瓢虫的寿命为3～6个星期，在整个生命周期中，一只瓢虫大约要吃掉5000只蚜虫。

Cotton has long replaced nettles as a source of fibre for clothing. Cotton now covers 2.5% of the world's cultivated land. It however uses 16% of the world's agricultural chemicals.

长久以来，棉花取代了荨麻，用于制作衣服。棉花的种植面积已经达到了全世界耕地的 2.5%，使用了全世界 16% 的农药。

Ladybirds are also called ladybugs, but they are neither bugs nor birds. In fact, they are a type of beetle.

瓢虫的英文名为 ladybird 或 ladybug，但瓢虫既不是鸟 (bird)，也不是臭虫 (bug)。事实上，瓢虫是甲虫的一个种类。

Think About It
想一想

Do you think that nettles, even though they sting, play an important role in our lives?

你认为扎人的荨麻在我们的生活中扮演着重要的角色吗?

如果让你穿上荨麻制成的牛仔裤,你会感觉怎样?

How would you like to be dressed in jeans made from nettles?

Would you enjoy living in a world without fashion, where you inherit your clothes from your grandparents?

你愿意生活在没有时尚的世界吗?在那里,你穿的是祖辈留下来的衣服。

你如何看待给其他物种以错误的命名?比方说,ladybird事实上既不是女士也不是小鸟。

What do you think of a language that calls something a name it is not, for instance a ladybird, which is neither a bird nor a lady?

Do It Yourself!
自己动手!

Look for a patch of nettles. Around Europe and North America you will find many. Take the time to make sure that there are no snakes or spiders. Now put on some gloves and pick a bag of nettles. Look under the leaves for ladybirds, moths, butterflies, larvae and aphids. See if you count at least ten different forms of life. When you get home wash the nettles and use them to make a soup. Do you need a recipe? Ask you mom to treat the nettles like spinach and let us know what it tastes like!

寻找一片荨麻。在欧洲和北美，你可以找到很多荨麻。要确保那里没有蛇或蜘蛛。然后带上手套，摘一袋荨麻。仔细观察荨麻叶子上的瓢虫、飞蛾、蝴蝶、昆虫的幼虫和蚜虫等。看看你是否能数出10种以上不同的生物。回家后，冲洗一下荨麻，然后把它们做成汤。需要食谱吗？问问妈妈如何烹饪菠菜的方法吧，然后告诉大家荨麻的味道哦！

TEACHER AND PARENT GUIDE

学科知识
Academic Knowledge

生物学	荨麻是蝴蝶、蛾、毛毛虫等幼虫的食物来源；野生或家养的大型牲畜不吃荨麻，因为荨麻带刺；瓢虫的自卫机制包括释放化学物质和保护色；如果食物短缺，瓢虫会吃掉自己软体的同胞；传统的农耕技术会在庄稼中留出一小块地方供昆虫来生存繁衍，以保障庄稼免于虫害。
化 学	荨麻富含维生素A、维生素C、铁、钾、锰和钙，是治疗风湿病的药物；荨麻多产说明土壤肥沃含氮；荨麻富含三种色素：叶绿素、叶黄素和胡萝卜素，可以用来制作冰淇淋或用于烘焙食品。
物 理	荨麻是中空的，里面充满了气体，为荨麻衣服提供了天然的保温效果；夏天时，纤维会聚集得很紧，保温效果降低。
工程学	荨麻纤维是白色丝状的，最长可达50毫米，比亚麻更加柔顺、细滑，是制作纺织物的理想原料；荨麻是一种天然的阻燃剂，可被用作石棉和墙面覆盖物的替代物。
经济学	棉花取代荨麻是由于其纤维丰富，但棉花种植产生了很多意想不到的后果，如大量的灌溉需求、农药的使用、农作物单一种植和转基因等问题；进口物品的选择标准不是质量和用途，而是时尚与消费者偏好。
伦理学	荨麻或许不被归为有机产品，却是在与生态和谐共生的环境中生长的。
历 史	世界上第一个荨麻织坊要追溯到青铜器时代的丹麦；伊丽莎白女王一世睡在荨麻床单上；拿破仑的军队穿着荨麻制成的衣服；第一次世界大战时，因为棉花受到限制，德国军队也是穿着荨麻衣服。
地 理	荨麻多见于欧洲、北美和亚洲，现在绝大多数的荨麻供给源于尼泊尔。
数 学	荨麻与棉花之间的资源效率对比计算。棉花需要农药种植，依赖进口，流行周期短，工人收入低；荨麻保存时间长，不需农药，可以世代流传。
生活方式	用"100%纯荨麻"或其他可再生、储量充足的纤维取代"100%纯（有机）棉"；欧洲传统用餐习惯喜欢将汤作为第一道菜，但现在不这样了。
社会学	瓢虫被叫做ladybird始于中世纪，那时候，农民向圣母玛利亚祈祷，寻求保护庄稼不受害虫侵扰，后来，一种保护庄稼的益虫出现了，人们就给它起名"ladybird"，以纪念圣母玛利亚。
心理学	妈妈的天性是保护、希望自己的孩子远离危险；享受父母与孩子共同进行实时实地的学习的机会。
系统论	荨麻促进生物多样性，保持土壤平衡。

教师与家长指南

情感智慧
Emotional Intelligence

瓢虫

瓢虫是一个细心的观察者,她注意到了小山羊非常饿。她忙碌而又精神集中,并没有将全部注意力从其食物——蚜虫上转移过来。瓢虫想知道是谁在教小山羊这些知识,想到应该是小山羊的妈妈,但是瓢虫也表达了这样的观点,那就是最好的老师也许应该是自己的亲身实践。而且,瓢虫为小山羊提供了时间和实地学习的经验,以及更广阔的视角——荨麻不仅仅是食物。瓢虫细致深入地讲述了关于生活方式的内容,分享了她所生活的生态系统中的作用方式。瓢虫表达了自己对小山羊情感上的认同。她将学习提升到了哲学层面,那就是我们要学会满足于当前自己所拥有的资源。

小山羊

小山羊饿极了,以致不能克制自己的情绪。他露出了一副很痛苦的表情。他做好了冒险尝试吃一次营养丰富的荨麻的准备,却换来了疼痛和不适。他坚信自己已经从妈妈那里学到了从健康到食物方面的大量知识,并且知道如何运用这些知识。小山羊意识到自己能力有限(比如做荨麻汤),但我们要知道小山羊还是虚心好学的。他求知欲很强而且很信任瓢虫。他想知道如何能简单应对现代时尚。小山羊还提出了更深刻的问题,人们为什么总在两种坏东西中做选择呢?这也为瓢虫阐述自己的处世哲学创造了氛围。

艺术
The Arts

荨麻富含黄绿到黄这一色调的色素。摘下一些荨麻叶子,放在水中煮沸。从锅里捞出煮过的荨麻,然后让水蒸发掉,这样就能收集到颜料了。当锅里的水蒸发到仅剩1厘米深时,你的颜料就收集好了,然后就可以作画啦!水蒸发得越多,颜色就越深。你可以把颜料混合到香草冰淇淋中,做出黄色或绿色的冰淇淋。

TEACHER AND PARENT GUIDE

思维拓展
Systems: Making the Connections

大自然不是孤立的系统，而是由很多相互联系的系统组成的一个复杂体系。生命是相互交织的网络。荨麻是牧场的组成部分，通常生长在牧草旺盛的地带。荨麻将这片牧草变成一个生命力非常旺盛的地带。荨麻不仅为蝴蝶、蛾等小生物提供生存之所，而且还通过展开协作来保护它的生殖繁衍不被饥饿的蚜虫所影响。这也解释荨麻何以发展成为瓢虫的避难所。高产的荨麻贡献着养料、纤维、彩色颜料、健康产品等甚至更多的东西，却仍然被称作杂草。人们已经忘记了，500年以前，几乎每一个身份显赫的人都是身穿荨麻制品，睡在荨麻制成的床单上。人们还忘记了荨麻所提供给我们的充足营养。非常高效、高产的荨麻生态系统，免费为我们提供了许多产品与服务。如果我们还认识不到自己本地现有资源的重要作用，不关心我们周围的生存环境，那么我们就还要依赖进口，而且搞不清这些舶来物是否对人体有害，是否危害着环境。我们的无知导致了很多机遇的丧失，比如促进本地就业、提高经济收入、提升能源利用效率等。种植荨麻不需农药，不需灌溉；加工荨麻也只需加工棉花或化纤产品所需原料的一小部分。是时候重新认识并利用我们周围现有的资源了。荨麻生命力旺盛，即便是经常收割，也能很快再长出来。我们需要增强发现并利用现有资源的能力，扭转资源长期短缺的困境，满足每个人的生存必需。

动手能力
Capacity to Implement

针织物不仅用于制作衣服，也用来制作书包、鞋、床品以及墙面装饰物。现在，我们要利用一种特殊纤维制成的针织物来设计一款书包。这种纤维在大自然中非常常见，不需人工种植，但却可以轻松收获。你的第一个任务是找到这种纤维，第二个任务是想出把它从纤维变成纤维织物的办法，比如机器加工或针织的方法，然后做成书包后卖出你的产品，让那些帮助你实现梦想的人健康幸福地生活。现在，你有行动方案了吗？

教师与家长指南

故事灵感来自

西比拉·索隆多
Sybilla Sorondo Myelzwynska

1963 年，西比拉·索隆多出生于美国纽约，17 岁时获得在伊夫圣罗兰设计工作室工作的机会。她致力于设计独特服饰，创立自己的品牌。在历经 20 多年的成功职业生涯后，西比拉决定休息一下，并出售了自己商业中的大部分收益。在思考了自己对地球、社会的责任后，她决心建立针织物自由联盟，旨在促进纺织业对可再生能源的利用以及确保原材料质量。西比拉的愿望是不仅制作漂亮舒适的衣服，还要让衣服背后承载一个美丽的故事。西比拉承诺，要消除服饰面料的有毒成分，寻找转基因棉花的替代品，减少对农药的依赖。她还承诺要改善生产流程，为工人创造更好的工作条件。西比拉要提供更多充满关爱、生态友好的纺织品。值得一提的是，西比拉烹制的荨麻汤非常美味，这些荨麻是她家周围牧场里种植的。

更多资讯

http://www.swicofil.com/products/016nettle.html

http://www.fabricsforfreedom.com/

图书在版编目（CIP）数据

带刺的荨麻：汉英对照／（比）鲍利著；田烁译．－－上海：学林出版社，2015.6
（冈特生态童书．第2辑）
ISBN 978-7-5486-0868-4

Ⅰ．①带… Ⅱ．①鲍… ②田… Ⅲ．①生态环境－环境保护－儿童读物－汉、英 Ⅳ．① X171.1-49

中国版本图书馆 CIP 数据核字 (2015) 第 086055 号

──────────────────────────────────

ⓒ 2015 Gunter Pauli
著作权合同登记号 图字 09-2015-446 号

冈特生态童书
带刺的荨麻

作　　者——	冈特・鲍利
译　　者——	田　烁
策　　划——	匡志强
责任编辑——	匡志强　蔡雩奇
装帧设计——	魏　来
出　　版——	上海世纪出版股份有限公司 学林出版社
	地　址：上海钦州南路 81 号　电话／传真：021-64515005
	网址：www.xuelinpress.com
发　　行——	上海世纪出版股份有限公司发行中心
	（上海福建中路 193 号　网址：www.ewen.co）
印　　刷——	上海图宇印刷有限公司
开　　本——	710×1020　1/16
印　　张——	2
字　　数——	5 万
版　　次——	2015 年 6 月第 1 版
	2015 年 6 月第 1 次印刷
书　　号——	ISBN 978-7-5486-0868-4/G・317
定　　价——	10.00 元

（如发生印刷、装订质量问题，读者可向工厂调换）